Learning Tr 1 2 3

Tools

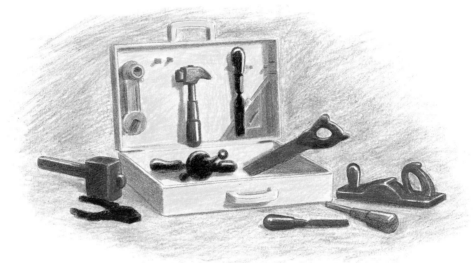

By Peter Stroud
Illustrated by Mike Lacey

Geddes + Grosset

Read this book and see if you can answer the questions.
Ask an adult or an older friend to tell you if your answers
are right or to help you if you find the questions difficult.
Often there is more than one answer to a question. Enjoy
getting to know about tools and learning how to use them.
But be careful, they can be very dangerous. Always ask
before you use tools.

First published in hardback 1991
Copyright © Cherrytree Press Ltd 1991

This paperback edition first published 1991 by
Geddes & Grosset Ltd
David Dale House
New Lanark ML11 9DJ

ISBN 1 85534 457 2

Printed and bound in Italy by L.E.G.O. s.p.a., Vicenza

All rights reserved. No part of this publication may be reproduced,
stored in a retrieval system, or transmitted, in any form or by any
means without the prior permission in writing of the publisher, nor
otherwise circulated in any form of binding or cover other than that in
which it is published and without a similar condition including this
condition being imposed on the subsequent purchaser.

Tools help us do things that we cannot do with our bare hands.
What do you use these tools for?

Once there were no tools.
Then people found out that they could break things with heavy stones.
They could cut things with sharp stones.
They could pierce things with pointed sticks.

Sticks and stones were the first tools.
Now most tools are made of metal and plastic.

Tools are not toys. Never play with them.
They are often sharp and dangerous.
Use them only when you have an adult to help.

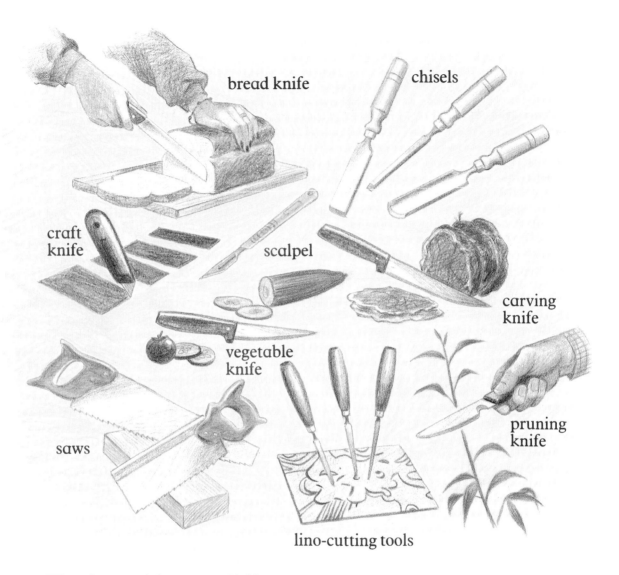

Each tool has a different job.
These tools are for cutting and shaping.
What other cutting tools do you know?

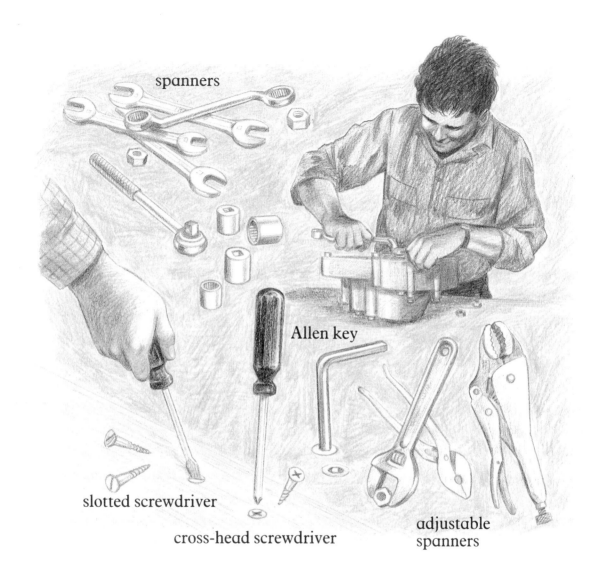

These tools help you turn things.
A screwdriver helps you drive a screw in.
A spanner helps you unscrew a nut.

You strike things with a hammer.
You can hammer a nail into a wall.
You can chop wood with an axe.
It strikes and cuts at the same time.

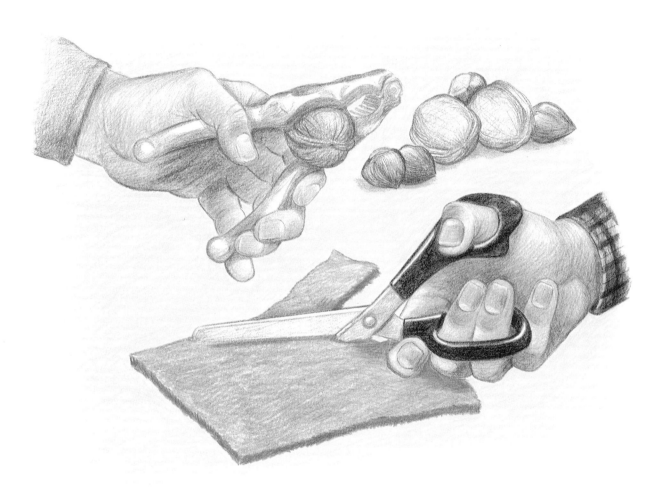

Some tools help you squeeze things.
Nutcrackers help you crack open nuts.
You cut things with scissors.
Can you see how they squeeze at the same time?

A ruler is a tool for measuring.
Can you think of other measuring tools?
How do they work?
What do they measure?

A comb keeps your hair free of tangles.
A toothbrush cleans your teeth.
What kinds of tools do you use for cleaning and tidying your home?

Take a look at your kitchen.
Find out what the tools are called.
What are they used for?
How do they work?

Can you find a tool that peels things?
Can you find a tool that rolls pastry?
Can you find one that opens cans and one that takes the tops off bottles?

Gardeners dig with a fork or spade.
They use their hands and feet to work them.
What tools do they use to cut plants?
What tools do they use to tidy up?

Carpenters build things with wood.
They have to measure, cut and join the wood.
This carpenter is making a bird-box.
What tools does he need?

wood saw

hacksaw

A carpenter uses a saw to cut wood.
Wood is quite soft. Metal is much harder.
A plumber uses a saw to cut a metal pipe.
But it is a different kind of saw.
It has smaller teeth. It is called a hacksaw.

Some tools can do several jobs.
A bricklayer joins bricks together with mortar.
He uses a builder's trowel to mix the mortar.
He uses it to carry the mortar, to spread it and to smooth it.

bricklayer's trowel

power saw

electric hedge clippers

electric drill

Working with hand tools is hard work.
The power comes from your muscles.
Power tools are much easier to use.
They use electric power.
Power tools and electricity are dangerous.
NEVER TOUCH THEM.

Some tools are not worked by hand.
They are part of machines.
They are called machine tools.
People use them in factories and workshops.
A lathe is a machine tool that is used for shaping wood and metal.

lathe

Draw a picture of a bird in a bird-box.
Ask an adult to help you make a frame for it.
What tools will you need?
What tools will you need to hang it up?

More about tools

The first tools
The first people to use tools lived in Africa more than a million years ago. They made simple tools from stone. They used their tools for killing animals and for cutting and hammering. The time when these people lived is called the Stone Age.

The first metal tools were made of bronze, a mixture of copper and tin. Bronze tools were used more than 5000 years ago during the Bronze Age.

The first iron tools were made about 3000 years ago during the Iron Age.

Flint and bone
Stone tools were often made of a kind of rock called flint. Flint is unusual because it is very hard but it can be chipped into flakes by another stone.

Sharp pieces of flint could be used as axe heads or as knives or scrapers. In the Stone Age, people made animal skins into clothes. They sewed them with needles made of bone.

Tools of the gods
The ancient Greeks knew how to shape metal. You heat it until it is hot enough to melt. Then a metalsmith can work it into shape. He holds the red-hot metal with tongs. Without tongs, he could not work the metal. How, the Greeks wondered, were the first tongs made, since there would have been no tongs to hold the hot metal? The gods, they thought, must have made the tongs and given them to the world.

All-purpose portable tool
Once people used feathers to write with. They had to sharpen the end of the quill to make the pen nib fine. People carried a folding pen-knife to cut their quill pens.

People soon realised that they could carry other small tools around in their pocket. Today, you can buy pen-knives that come with several blades as well as files, corkscrews, bottle openers, screwdrivers, scissors, toothpicks, hooks and pins. The tools all fold neatly back into the case.

1

1 What are tools?

2 What kind of tool would you use to slice a loaf of bread?

3 What do scissors do?

4 What can you do with a hammer?

5 Which of these is the odd one out: hammer, spanner, saucer, fork?

6 What are these tools called?

2

7 What is the difference between a shovel and a spade?

8 What are most tools made of?

9 What were the first tools made of?

10 In your kitchen you may find a sieve, a colander and a strainer? What are they used for? What is the difference between them?

11 Can you name five different cutting tools?

12 What tool would you use to cut a metal pipe?

13 What tool would you use to cut down a tree? What tool would you use to chop logs?

14 Who would use a hoe? A carpenter, a cook or a gardener?

3

15 What tools would you need to make a folder? Gather your tools and make one. Write the answers to these questions in a notebook. Make a note of any questions you want to ask. Make notes about tools and their uses. Draw pictures of them.

16 Tools, implements and machines are different things we use. Find out the meaning of each word.

17 What tools do you use each day? Make a list of them in the order in which you use them. Your toothbrush will probably come first.

18 A plane is a tool used for smoothing wood. Find out why it is called a plane.

19 What is a vice?

20 What are these tools used for – rasp, awl, sickle, gimlet?

21 Where would you find a pitchfork? What is it used for?

22 How did jigsaw puzzles get their name?

23 The first hammer and the first knife were simply bits of rock. Knives and hammers have changed a lot. Choose a particular kind of tool and see how it has changed since it was first invented. Go to the library or to a museum and see if you can find pictures of the tool being used.

24 Do animals use tools?

25 Make a list of different kinds of brush. Say what they are made of, what they are used for and what they look like.

Index

adjustable spanners 7
Africa 21
Allen key 7
ancient Greeks 21
axe 8, 21
bird-box 15, 20
blades 21
bone tools 21
bottle opener 13, 21
bread knife 6
bricklayer 17
Bronze Age 21
bronze tools 21
builder's trowel 17
can opener 13
carpenter 15, 16
carpentry tools 15, 16
carving knife 6
chisels 6
chopping 8
cleaning tools 11
comb 11
corkscrew 21
craft knife 6
cross-head screwdriver 7
cutting tools 6, 8, 9, 15, 16
danger 5, 18
drill 18
electric drill 18
electric hedge clippers 18
electricity 18
electric mixer 18

files 21
first tools 4, 21
flint tools 21
fork 14
frame 20
gardeners 14
gardening tools 14
hacksaw 16
hammer 3, 8
hedge clippers 18
hooks 21
Iron Age 21
iron tools 21
joining wood 15
kitchen tools 12, 13
knives 21
lathe 19
lino-cutting tools 6
machine tools 19
measuring tools 10, 15
metal 5, 16, 19, 21
metalsmith 21
mixer 18
mortar 17
nail 8
needles 21
nut 7
nutcrackers 9
pen-knife 21
pins 21
plastic 5
plumber 16

power saw 18
power tools 18
pruning knife 6
rake 3
ruler 10
saws 3, 6, 16, 18
scalpel 6
scissors 3, 9, 21
scrapers 21
screw 7
screwdrivers 3, 7, 21
shaping tools 6, 19
slotted screwdriver 7
spade 3, 14
spanners 7
spoon 3
squeezing tools 9
sticks 4, 5
Stone Age 21
stone tools 4, 5, 21
striking tools 8
tong 21
toothbrush 11
toothpicks 21
trowel 17
turning tools 7
vegetable knife 6
whisk 3
wood 8, 15, 16, 19
wooden spoon 3
vegetable knife 6